## FREAKY TRUE SCIENCE

# FREAKY STORIES ABOUT TECHNOLOGY

BY RYAN NAGELHOUT

Gareth Stevens
PUBLISHING

Please visit our website, www.garethstevens.com. For a free color catalog of all our high-quality books, call toll free 1-800-542-2595 or fax 1-877-542-2596.

Library of Congress Cataloging-in-Publication Data

Names: Nagelhout, Ryan, author.
Title: Freaky stories about technology / Ryan Nagelhout.
Other titles: Freaky true science.
Description: New York : Gareth Stevens Publishing, [2017] | Series: Freaky true science | Includes bibliographical references and index.
Identifiers: LCCN 2016003381 | ISBN 9781482448382 (pbk.) | ISBN 9781482448443 (library bound) | ISBN 9781482448429 (6 pack)
Subjects: LCSH: Technology–Miscellanea–Juvenile literature.
Classification: LCC T48 .N34 2017 | DDC 602–dc23
LC record available at http://lccn.loc.gov/2016003381

First Edition

Published in 2017 by
**Gareth Stevens Publishing**
111 East 14th Street, Suite 349
New York, NY 10003

Copyright © 2017 Gareth Stevens Publishing

Designer: Sarah Liddell
Editor: Ryan Nagelhout

Photo credits: Cover, p. 1 (artificial heart) Danor Aharon/Shutterstock.com; cover, p. 1 (circuit board used throughout book) pzAxe/Shutterstock.com; cover, p. 1 (mouse) Palau/Shutterstock.com; background throughout book jeka84/Shutterstock.com; pp. 5, 7, 9, 11, 13, 15, 17, 19, 21, 23, 25, 27, 29 (hand used throughout) Helena Ohman/Shutterstock.com; pp. 5, 7, 9, 11, 13, 15, 17, 19, 21, 23, 25, 27, 29 (texture throughout) Alex Gontar/Shutterstock.com; p. 4 Mik81/Wikimedia Commons; p. 5 orangecrush/Shutterstock.com; p. 7 Hovev/Wikimedia Commons; p. 9 Monty Rakusen/Cultura/Getty Images; pp. 10, 13 ChinaFotoPress/Contributor/Visual China Group/Getty Images; pp. 11, 29 Bloomberg/Contributor/Bloomberg/Getty Images; p. 15 (cube houses) Igor Plotnikov/Shutterstock.com; p. 15 (Ryugyong Hotel) Harpsichord246/Wikimedia Commons; p. 15 (Capital Gate Tower)Zhukov Oleg/Shutterstock.com; p. 15 (CCTV Tower) axz700/Shutterstock.com; p. 17 Luis Lamar/National Geographic/Getty Images; p. 19 Satyrenko/Shutterstock.com; p. 21 (flying drone) Chris McKay/Contributor/Getty Images Entertainment/Getty Images; p. 21 (robotic lawn mower) Education Images/Contributor/Universal Images Group/Getty Images; p. 23 (cocoa pods) WIBOON WIRATTHANAPHAN/Shutterstock.com; p. 23 (*Crinipellis perniciosa* mushroom) Luigi Chiesa/Wikimedia Commons; p. 25 (main) NASA/Handout/Getty Images News/Getty Images; p. 25 (lunch tray) Sjschen/Wikimedia Commons; p. 27 ROBYN BECK/Staff/AFP/Getty Images.

All rights reserved. No part of this book may be reproduced in any form without permission in writing from the publisher, except by a reviewer.

Printed in the United States of America

CPSIA compliance information: Batch #CS16GS: For further information contact Gareth Stevens, New York, New York at 1-800-542-2595.

# CONTENTS

Science at Work ................................................ 4
Stuxnet Strikes ................................................. 6
3-D Printing ..................................................... 8
Clones That Save ............................................ 12
China's Building Boom .................................. 14
Wearing Technology ...................................... 16
Cleaning Light ............................................... 18
Robots at Work .............................................. 20
Chocolate Panic ............................................. 22
Food in Space ................................................ 24
Not So Smart ................................................. 26
The Future of Tech ........................................ 28
Glossary ........................................................ 30
For More Information .................................. 31
Index ............................................................. 32

Words in the glossary appear in **bold** type
the first time they are used in the text.

# SCIENCE AT WORK

Today you might think of technology as just computers and cell phones, but the term covers all kinds of amazing things. Technology is anything related to science that is used to solve a problem. It can make existing products or tools better or present something new that completely changes the way we live. Even the way we design buildings, called architecture, has an element of technology in it.

Sometimes new technology can seem a bit strange. Eventually, we get used to having refrigerators in our kitchens and phones in our pockets. But there's some technology working around you right now that you just don't know about. Some of it is amazing, and some of it has effects its creators never meant. That's when technology can get just plain freaky.

PROJECTION KEYBOARD

# THE ANCIENT INCA

Some technology is freaky because it seems very modern despite being very old. The Inca were a native civilization that lived in South America until the 16th century. Despite living hundreds of years ago, the Inca managed to build a **network** of roads about 24,000 miles (38,600 km) long. They also kept their own calendar, built homes with advanced architecture, and made amazing woven rope bridges that European explorers were afraid to cross.

THE INCA ROAD NETWORK EVEN REACHED THE ANCIENT CITY OF MACHU PICCHU, WHICH WAS BUILT HIGH IN THE ANDES MOUNTAINS.

# STUXNET STRIKES

The US military is the most powerful in the world. But thanks to some impressive technology and smart **hackers**, modern governments can attack without firing a single gun. Starting in 2009, computer experts in the United States and Israel created a computer virus designed to ruin computers running Iran's nuclear weapons program.

Called "Stuxnet," the virus spread to hundreds of thousands of different machines all over the world, but researchers tracked down its target: Iran. After years of study, scientists learned the **complex**, professionally made computer virus Stuxnet helped destroy thousands of **centrifuges** in Iran, slowing down their weapons making without dropping a single bomb.

Stuxnet is just one of the freaky things technology is capable of. How much do you really know about the technology working all around us?

**FREAKY FACTS!**

In North Korea, the police must give you permission to own a computer. Many websites are blocked, and only a few people can reach the "open" Internet.

STUXNET DID DAMAGE TO MANY DIFFERENT SYSTEMS AND COMPUTERS AROUND THE WORLD, BUT ITS TARGET WAS NUCLEAR FACILITIES IN IRAN.

NATANZ NUCLEAR FACILITY IN IRAN

## FAILURE IN NORTH KOREA

In 2015, the news group Reuters reported the United States had tried a similar attack on North Korea's nuclear program around the same time Stuxnet struck Iran. Reuters sources said Stuxnet developers created a similar virus that would act when it found language in Korean on an infected computer. North Korea, however, is cut off from many computer networks, and the agents had trouble getting to the main machines that run North Korea's nuclear program.

# 3-D PRINTING

You've probably printed an essay out on paper before, but have you ever printed something made of plastic? The technology to do that—called 3-D printing—uses high-tech computers to design and then "print" an object in layers using many different materials.

Most 3-D printers use plastic or some other strong material to make any number of objects. If you can use a 3-D design program to make it, a 3-D printer can print it out. People have used 3-D printers to make models, toys, jewelry, or even replacement parts for tools. Lots of different websites offer 3-D print designs that people can download for themselves. Anyone can find a 3-D printing design online, send it to a printer, and have that object created in a few hours.

**Many different materials can be used to 3-D print, including plastic, porcelain, and metals like gold, silver, or aluminum. Some 3-D printers can even use wax!**

## CAN YOU PRINT TOO MUCH?

Some worry that the risks of 3-D printing may outweigh its exciting possibilities. A few politicians and scientists worry people will use 3-D printers to make dangerous objects. For example, someone may make parts of guns or other objects that can do harm to a lot of people without having to buy these dangerous tools legally. The law often has to catch up to technology when it can put people at risk.

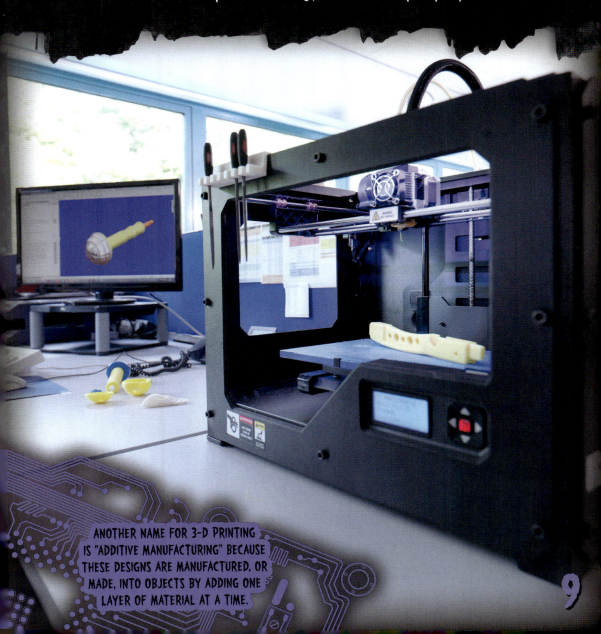

ANOTHER NAME FOR 3-D PRINTING IS "ADDITIVE MANUFACTURING" BECAUSE THESE DESIGNS ARE MANUFACTURED, OR MADE, INTO OBJECTS BY ADDING ONE LAYER OF MATERIAL AT A TIME.

3-D PRINTED VERTEBRAE

Some replacement body parts—like hip joints—are already being made with 3-D printers! Companies can also use them to make replacement bones using a **polymer**. But some scientists are trying to make 3-D printers that use organic material to print living objects. They think these "bio-printers" could even be used to create human **organs**! Researchers at Johns Hopkins and Princeton Universities tested a 3-D printed ear, which they showed off in 2015.

Scientists—and companies hoping to sell these organs—hope to use live cells to make human tissues one layer at a time. These tissues can be used to help people fix damaged organs. Some scientists, however, are working on printing more complex materials in a gel, which could lead to printing entire human organs.

FREAKY FACTS!

In 2014, printed body parts made companies more than $500 million, a 30 percent increase from the year before. Companies expect that number to rise as the technology improves.

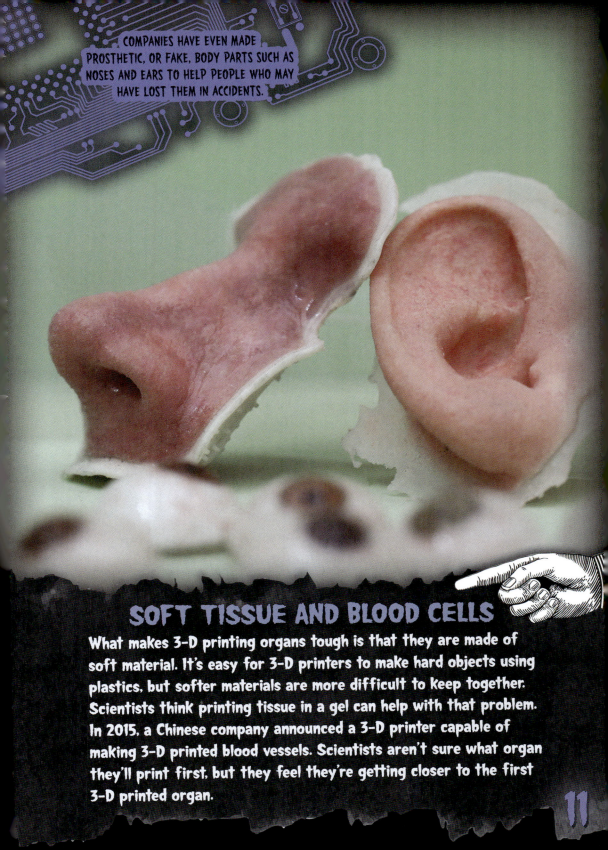

COMPANIES HAVE EVEN MADE PROSTHETIC, OR FAKE, BODY PARTS SUCH AS NOSES AND EARS TO HELP PEOPLE WHO MAY HAVE LOST THEM IN ACCIDENTS.

## SOFT TISSUE AND BLOOD CELLS

What makes 3-D printing organs tough is that they are made of soft material. It's easy for 3-D printers to make hard objects using plastics, but softer materials are more difficult to keep together. Scientists think printing tissue in a gel can help with that problem. In 2015, a Chinese company announced a 3-D printer capable of making 3-D printed blood vessels. Scientists aren't sure what organ they'll print first, but they feel they're getting closer to the first 3-D printed organ.

# CLONES THAT SAVE

A clone is an exact copy of a living thing. While scientists claim human cloning is still not possible, they've cloned some other animals, such as sheep. Since Dolly the sheep was cloned in 1996, scientists have managed to copy animals such as cattle, a cat, a deer, a rabbit, a horse, and even a monkey.

Scientists hope they can use cloning to save endangered species, or animals close to dying out altogether. In 2001, an endangered species of ox called a guar was cloned. It only survived a few days, but in 2003, a type of ox called the banteng was successfully cloned. Scientists hope they can use DNA to clone larger mammals, such as the northern white rhinoceros, even after the last of their kind dies out!

## FREAKY FACTS!

Cloned animals don't always look exactly like the animal they were cloned from. The first cat to be cloned, Cc, was a female calico that looked very different from her mother!

# CLONING PROBLEMS

Scientists are excited about its possibilities, but realize cloning isn't perfect. It's difficult to create clones, and many don't live long, healthy lives. Cloned animals often have weaker immune systems than ordinary animals and may age faster. Scientists also worry that cloning endangered species from a few remaining animals may not create enough genetic diversity, or differences, to keep the population healthy in the wild.

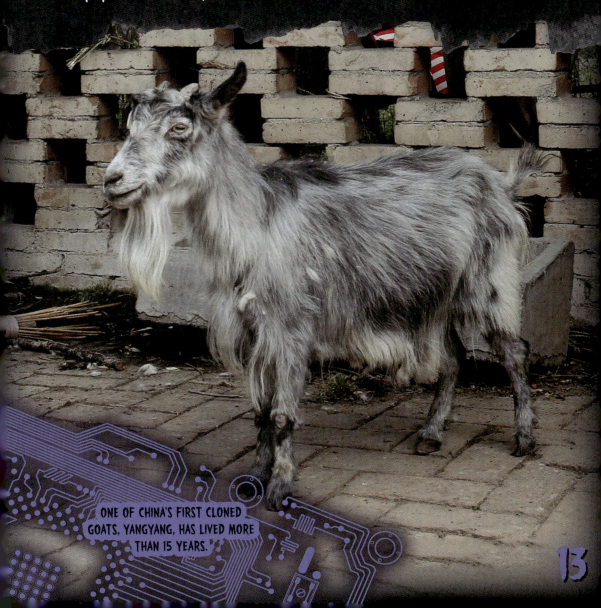

ONE OF CHINA'S FIRST CLONED GOATS, YANGYANG, HAS LIVED MORE THAN 15 YEARS.

# CHINA'S BUILDING BOOM

Some of the wildest architecture on Earth is built in China. Large parts of the population are moving to big cities in China, including a growing wealthy class that helps pay for the construction of massive skyscrapers. Architects let their imaginations run wild! For example, the China Central Television building in Beijing looks like a **tripod** that's missing a leg. A number of buildings in Beijing look like eggs.

Elsewhere, the Tianzi Hotel in Hebei looks like three standing ancient Chinese gods. The city of Wuxi has a building shaped like a teapot with a hole in it. But the weirdest might be in Huainan, where architecture students designed a building shaped like a piano and violin!

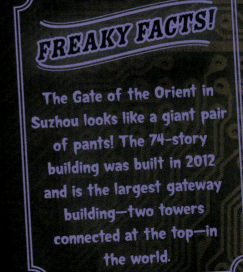

**FREAKY FACTS!**

The Gate of the Orient in Suzhou looks like a giant pair of pants! The 74-story building was built in 2012 and is the largest gateway building—two towers connected at the top—in the world.

CHINA ISN'T THE ONLY COUNTRY WITH ARCHITECTS DESIGNING FREAKY BUILDINGS. THERE ARE PLENTY OF WILD CREATIONS ALL OVER THE WORLD.

# FREAKY BUILDINGS AROUND THE GLOBE

**CHINA CENTRAL TELEVISION BUILDING**
BEIJING, CHINA

**CAPITAL GATE TOWER**
ABU DHABI, UNITED ARAB EMIRATES

**RYUGYONG HOTEL**
PYONGYANG, NORTH KOREA

**CUBE HOUSES**
ROTTERDAM, NETHERLANDS

## THE END OF WEIRD?

In Wujin, a series of buildings that look like lotus flowers was built in 2013. In Fushun, city planners built a giant metal ring to attract tourists. But in 2014, Chinese president Xi Jinping said art should serve the people and that it should "be like sunshine from the blue sky and the breeze in spring that will inspire minds, warm hearts, **cultivate** taste, and clean up **undesirable** work styles." Many think this means the end of China's unusual-looking buildings.

15

# WEARING TECHNOLOGY

You might know someone who wears a smart watch or fitness tracker on their wrist. This type of wearable technology can track your heart rate, count your steps, or just tell you when your phone has a new text message. Scientists are working on all kinds of wearable technology, like the Archelis. Made in Japan, it's a wearable chair designed for surgeons so they can sit while performing long operations.

Other freaky wearable tech includes an exoskeleton, which is a structure worn outside the body that helps give it strength. Most exoskeletons scientists are working on today are meant to help injured people walk or use other limbs again. Newer designs, however, can help healthy humans lift up a car!

**FREAKY FACTS!**

New exoskeleton designs for military use would give soldiers superhuman powers. The TALOS "Iron Man suit" is bulletproof and lets soldiers carry heavy weapons while giving them special information in a HUD, or heads-up-display, in its helmet.

# BIOHACKERS

Some people take wearing technology a bit too far for most—they put it under their skin! These "biohackers" implant magnet-based LEDs under their skin. The device, called the Northstar V1, is about the size of a coin. When placed under the skin, a magnet is used to make the LEDs flash. The group implanting the LEDs says they were inspired by bioluminescent organisms, or plants and animals that naturally create light.

THE EXOSUIT WAS DESIGNED LIKE A ONE-PERSON SUBMARINE THAT LETS PEOPLE DIVE DEEP UNDERWATER.

# CLEANING LIGHT

Ultraviolet, or UV, light can be used by scientists for some amazing things. UV light causes ionization, which is when energy is transferred to an object to give it a particular charge. Usually one or multiple electrons are exchanged between atoms during ionization. When scientists make their own UV light, however, they can even use it to "clean."

When UV light is made artificially, it can be used to **disinfect** different things. Some systems use UV light to make water safe for drinking. The UV light affects the nucleus of bacteria, destroying it or harming its DNA so it can't reproduce and spread to make someone sick. Ultraviolet light can be used to clean hospital equipment and to treat wastewater before it flows into waterways.

## FREAKY FACTS!

Ultraviolet light is also used to study space. The hotter something is, the more UV light it gives off. Scientists use special cameras to take UV pictures of objects in space to learn more about their temperature.

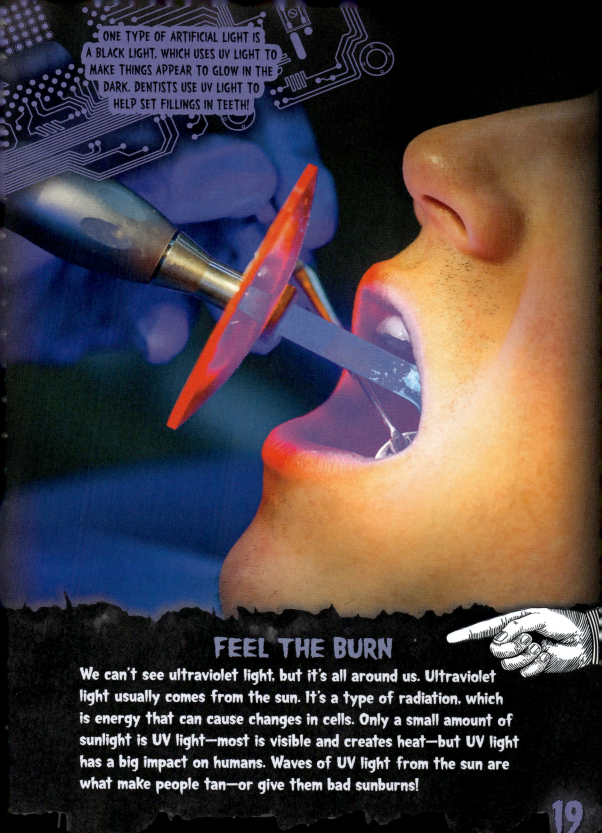

ONE TYPE OF ARTIFICIAL LIGHT IS A BLACK LIGHT, WHICH USES UV LIGHT TO MAKE THINGS APPEAR TO GLOW IN THE DARK. DENTISTS USE UV LIGHT TO HELP SET FILLINGS IN TEETH!

## FEEL THE BURN

We can't see ultraviolet light, but it's all around us. Ultraviolet light usually comes from the sun. It's a type of radiation, which is energy that can cause changes in cells. Only a small amount of sunlight is UV light—most is visible and creates heat—but UV light has a big impact on humans. Waves of UV light from the sun are what make people tan—or give them bad sunburns!

# ROBOTS AT WORK

Your parents might have a cool robot that works as a vacuum, keeping your house clean. But do they have one cutting your lawn? A number of different companies sell robotic lawn mowers. They're a bit expensive, but work just like cleaning robots, except they have sharp blades to cut grass.

Robotic lawn mowers are a type of drone, a robot controlled through radio signals instead of being piloted directly by a human. There are drones that fly, swim, and even do your chores. The cutting drones can work constantly, finding parts of your lawn that are too tall at all hours of the day. Some can be controlled with a cell phone and go to their charging station when their batteries are low.

**FREAKY FACTS!**

Many robotic lawn mowers are much quieter than gas-powered, push, or riding lawn mowers. They also use much smaller blades, and some can work in the rain!

AS DRONES GET CHEAPER AND EASIER TO USE, MORE PEOPLE ARE ADDING ROBOTS TO THEIR EVERYDAY LIVES. WOULD YOU WANT TO FLY A DRONE?

FLYING DRONE

ROBOTIC LAWN MOWER

## DANGEROUS DRONES?

The rise of flying drones, also called unmanned aircraft systems, has created some interesting problems in different areas. Drones controlled by people flying them in their backyard have ended up in some strange places. In 2015, for example, a drone crashed on the lawn of the White House in Washington, DC. The city banned drone flights earlier that year. Flying drones that weigh a certain amount now have to be registered with the Federal Aviation Administration (FAA).

# CHOCOLATE PANIC

Can you imagine a world without chocolate? It almost happened when cacao trees—the trees that produce the seeds used to make chocolate—started dying from a fungus in the late 1980s. Called "witches' broom fungus," it started killing massive amounts of cacao trees in Brazil. Scientists worried the fungus could spread to other cacao-growing countries and maybe even kill the tree off entirely.

The chocolate company Mars—worried that chocolate prices could go through the roof if the fungus spread—gave $10 million to a fund designed to **sequence** the tree's genome. In 2010, the entire genome project's findings were put on a website. Scientists think they can use the genome to make trees that are more resistant to witches' broom fungus.

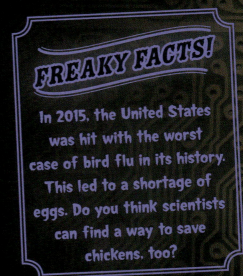

**FREAKY FACTS!**

In 2015, the United States was hit with the worst case of bird flu in its history. This led to a shortage of eggs. Do you think scientists can find a way to save chickens, too?

# BYE-BYE, BANANAS?

In the 1950s, Panama disease affected banana growers in Central and South America. The Gros Michel banana, the most flavorful known to man, basically became extinct. Banana growers had to pick another banana to sell and chose the Cavendish. Forty years later, however, a new strain of the fungus that harmed Gros Michel bananas started to kill Cavendish bananas. Many worry that unless scientists come up with a plan to fight the fungus, we'll lose Cavendish bananas, too!

COCOA PODS

CRINIPELLIS PERNICIOSA MUSHROOM, WHICH CAUSES WITCHES' BROOM FUNGUS

FUNGUS CAN CAUSE PROBLEMS FOR MANY DIFFERENT PLANTS, BUT GENOME SEQUENCING MEANS THE PLANT'S DNA CAN BE CHANGED TO RESIST ITS IMPACT.

# FOOD IN SPACE

Astronauts are just like us: They have to eat! They also have to bring anything they eat with them when they go into space. But they're not exactly grilling steaks or mixing salads aboard the International Space Station (ISS). The National Aeronautics and Space Administration (NASA) designs special meals for astronauts to take with them into space.

Most of the food is dehydrated, which means the water is taken out. It's added again just before the food is eaten. This allows the food to last a long time before going bad. It also weighs less, which is important when you're leaving Earth! Most food is thermostabilized, or heated to a high level to kill things that cause food to spoil. A few shelf-stable foods, however, are taken into space without any changes.

## FREAKY FACTS!

John Glenn was the first American to eat in space. He ate applesauce from a tube aboard *Friendship 7* in 1962. The experiment helped prove people could eat and digest food in space.

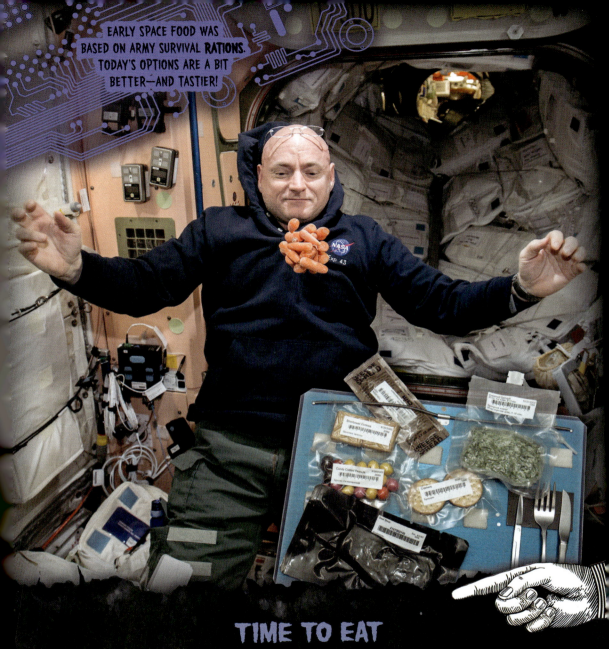

EARLY SPACE FOOD WAS BASED ON ARMY SURVIVAL **RATIONS**. TODAY'S OPTIONS ARE A BIT BETTER—AND TASTIER!

## TIME TO EAT

There are three kinds of food on board the ISS. The daily menu is picked by astronauts. Extravehicular Activity (EVA) food is used when astronauts are working outside the ISS and miss normal meals. Safe Haven food is used only in case of an emergency on board. Before going into space, they sample the menu options and pick their favorites. Food comes in single-serving containers to limit mess.

25

the temperature in their home from anywhere they can get an Internet or phone connection.

But not all smart technology is very smart. In 2015, a company announced a smart belt for people to wear around their waists. Belty connects to your phone and can sense when you've had too much to eat. When your waist gets bigger, the belt loosens up a bit. Would you rather adjust your belt, or have it adjust to you?

### FREAKY FACTS!

Each year, companies announce countless crazy pieces of technology at the Consumer Electronics Show (CES) in Las Vegas, Nevada. That's where most people get their first look at the freaky future.

## SMART OR SILLY?

Some "smart" technology you might find a bit freaky includes a water bottle that reminds you when to drink and a spoon with holes in it that get bigger if you're eating too "fast." There's even a "smart" bookmark that figures out what page it's saving. It then sends the page number to your phone in a text message when you're ready to read again. Many people wonder why we need these smart devices to do things plenty of "dumb" technology can already manage.

DO YOU NEED A SMART BELT TO KEEP YOUR PANTS UP? WHAT ABOUT A TOUCH SCREEN IN THE DOOR OF YOUR FRIDGE? SOME TECHNOLOGY IS DOWNRIGHT FREAKY.

# THE FUTURE OF TECH

If you've ever watched an old science fiction movie that imagined the future, you know people have some pretty wild ideas. We may not have flying cars and jet packs, but some modern technology would have shocked people a few decades ago.

Japan, for example, has a hotel almost entirely staffed by robots. A dinosaur robot speaks English and helps people check in, and other humanlike robots find information on touch-screen panels. Some cars can park themselves with the press of a button. Other companies—like technology company Google or transportation company Uber—hope to develop self-driving cars. With drones flying overhead and cloned animals living among us, maybe the freaky future is a lot closer than we think.

**FREAKY FACTS!**

Concorde jets were some of the fastest commercial airplanes ever built. But flying the planes long distances was the only way to make money with them. Sometimes making money is more important than building the best technology. The last Concorde jet flew in 2003.

JAPAN'S ROBOT HOTEL HAS ALL KINDS OF DIFFERENT ROBOTS ON STAFF. WOULD YOU LIKE A RAPTOR TO CHECK YOU IN?

## WHY NO JET PACKS?

A company called Bell Aerosystems tested a jet pack during the 1960s. So why don't we have them today? The weight of fuel and the safety of having a rocket strapped to your back are just a few reasons why jet packs never took off. Even jet packs made today only have enough fuel for about 30 seconds of flying time, which isn't very useful. Even if it looks really cool, there are just easier—and safer—ways for people to get around.

# GLOSSARY

**centrifuge:** a machine used to separate fluids using rotating force

**complex:** made up of many parts

**cultivate:** to develop by careful attention

**digest:** to break down food inside the body so the body can use it

**disinfect:** to kill or remove the ability of bacteria to spread

**hacker:** an expert at solving problems with or taking over computers

**network:** a system of related or connected parts

**organ:** a part inside an animal's body

**polymer:** a chemical compound made up of repeating structural units

**rations:** food or provisions

**sensor:** a device that senses information and sends it out for use

**sequence:** a series of connected events

**tripod:** something resting on three legs

**undesirable:** not wanted

# FOR MORE INFORMATION

## BOOKS

Barnham, Kay. *Could a Robot Make My Dinner? And Other Questions About Technology.* Chicago, IL: Capstone Raintree, 2014.

Marsico, Katie. *Drones.* New York, NY: Children's Press, 2016.

O'Neill, Terence, and Josh Williams. *3D Printing.* Ann Arbor, MI: Cherry Lake Publishing, 2013.

## WEBSITES

**Cacao Genome Database**
*cacaogenomedb.org*
Find more information about the chocolate plant genome project on this site.

**Food for Space Flight**
*nasa.gov/audience/forstudents/postsecondary/features/F_Food_for_Space_Flight.html*
Read more about the kinds of food astronauts have had in space here.

**Makerbot Thingiverse**
*makerbot.com/thingiverse*
Find free 3-D printing designs you can print out and use here!

**Publisher's note to educators and parents:** Our editors have carefully reviewed these websites to ensure that they are suitable for students. Many websites change frequently, however, and we cannot guarantee that a site's future contents will continue to meet our high standards of quality and educational value. Be advised that students should be closely supervised whenever they access the Internet.

# INDEX

3-D printing  8, 9, 10, 11
Archelis  16
architecture  14, 15
astronauts  24, 25
bananas  23
biohackers  17
cacao trees  22
China  14, 15
chocolate  22
cleaning robots  20
clones  12, 13
computer virus  6, 7
dehydrated food  24, 25
DNA  12
Dolly the sheep  12
drones  20, 21, 28
endangered species  12, 13
exoskeleton  16
ionization  18
jet packs  28, 29
Panama disease  23
replacement body parts  10, 11
robotic lawn mowers  20
robots  20, 28
self-driving cars  28
smart belt  26, 27
"smart" technology  26, 27
Stuxnet  6, 7
TALOS  16
UV light  18, 19
wearable technology  16, 17
witches' broom fungus  22